THE HAIR SCALE
IDENTIFICATION GUIDE TO

Terrestrial Mammalian
Carnivores of Canada

THE HAIR SCALE
IDENTIFICATION GUIDE TO
Terrestrial Mammalian Carnivores of Canada

JUSTIN KESTLER

PELAGIC PUBLISHING

Published by Pelagic Publishing
20–22 Wenlock Road
London N1 7GU,
UK

www.pelagicpublishing.com

*The Hair Scale Identification Guide
to Terrestrial Mammalian Carnivores of Canada*

Copyright © 2022 Justin Kestler

A CIP record for this book is available from the British Library

ISBN 978-1-78427-282-1 Pbk
ISBN 978-1-78427-283-8 ePub
ISBN 978-1-78427-284-5 PDF

https://doi.org/10.53061/YGXZ6719

All photographs © Justin Kestler unless mentioned below

Cover images: Grizzly bear in a river on a rainy day, Bella Coola, British Columbia, Canada © iStock/Paul Sahota; grizzly bear hair 50×; grizzly bear hair 200×.

pp.14–15: Red fox in countryside © Adobe Stock/Mark Medcalf. pp.28–29: Canadian lynx in the wild © Adobe Stock/Jillian. pp.36–37: Striped skunk (*Mephitis mephitis*) looks down side of log © Adobe Stock/Geoff Kuchera. pp.42–43: American marten or American pine marten (*Martes americana*) © Adobe Stock/Juan Carlos Munoz. pp.64–65: Raccoon © Adobe Stock/ Carol Hamilton. pp.68–69: King of the North © Adobe Stock/Zahi.

Design and typesetting by Christopher Bromley
Printed in England by Swallowtail

MIX
Paper from
responsible sources
FSC® C113523
www.fsc.org

This guide is dedicated to the memory of Teri Winter –
friend, teacher, mentor and wonderful human being
who inspired many.

Contents

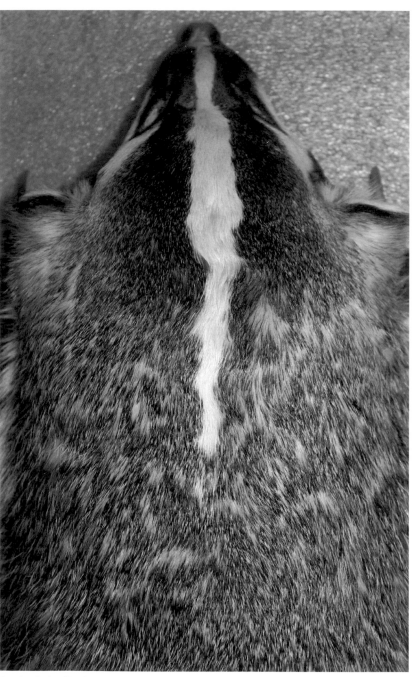

American Badger (*Taxidea taxus*)

Acknowledgements

This study was made possible thanks to the encouragement and effort of the dedicated Natural Environment & Outdoor Studies (NEOS) Faculty at Sault College. A big thank you goes out to Sault College's Teri Winter, Elisa Muto, Chad Ritesma, Rob Routledge, Lynn Goulding, Maxime Gerin-Ouellet, as well as Dr. Burton Lim and Jacquline Miller from the Royal Ontario Museum, and the Calgary Zoo, including both Alyssa Friesen and Allison Scovil, who all gave considerable time out of their busy work schedules to assist with this project. A special thank you to Alexander Keszei, whose talents and time put into the layout of the guide cannot be overstated. Finally, Teri Winter's incredible mentorship should also be noted, as without it this reference guide would not exist.

A. S. Adorjan & G. B. Kolenosky's seminal 1969 work *A Manual for the Identification of Hair of Selected Ontario Mammals* has been a great asset and source of inspiration as well.

I would also be remiss if I didn't thank my family, who have supported me throughout my journey. Their patience, perspective, and unconditional love have been the backbone to my budding naturalist career.

Happy outdoor exploring to all.

Introduction and Guide Interpretation

This guide is intended to be used as a reference for hair samples found in the field which require further identification via light microscope in a laboratory or home setting. All hair sample specimens were provided by one of the following: Sault College, the Royal Ontario Museum (ROM) and the Calgary Zoo.

All hair scale samples found in this reference guide are organized alphabetically under the order Carnivora – family first (Canidae, Felidae, Mephitidae, Mustelidae, Procyonidae, Ursidae) and subsequently organized from the smallest to the largest species within their family. The purpose is to facilitate a quick and structured way of searching for the intended hair scale specimen. The names and taxonomy are based on the comprehensive *The Natural History of Canadian Mammals* resource (Naughton 2012).

The similar species (overall characteristics) category takes into account:

    1) the scale type,

    2) the gross description,

    3) the key hair scale characteristics,

    4) and the overlapping range of each species.

Where range and (most of) the three other aspects converge between two carnivores, relevant similar species will be listed.

Accompanying images for each species represented in the guide are taken from the top (the back and/or withers) of the animal. Where necessary, the undercoat and/or other pelage distinctions are provided.

To maintain consistency, only the *guard hairs* of each mammal specimen were used. This is because guard hairs are easier to handle and much easier obtain a hair scale impression of. Additionally, only the magnified *medial* portion of the hair is shown in this guide. The magnification level for each sample hair specimen is depicted at 50× and 200× respectively. The goal of this guide is to aid seasoned researchers and amateurs alike who conduct mammalian carnivore fieldwork across Canada.

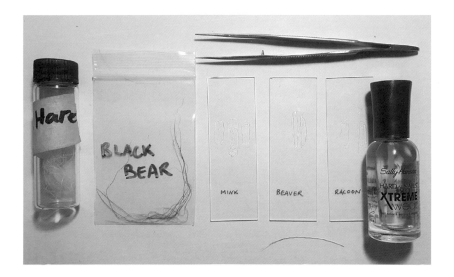

Fig. 1: Image of materials used for imprinting hair pattern for imaging.

Fig. 2: Close-up image of hair impression left in nail polish.

MATERIALS AND METHODS

An array of taxidermy and pelt specimens retained by Sault College, the Royal Ontario Museum and the Calgary Zoo were sampled for hair identification. Each species was categorized accordingly. In order to ensure that there was no cross-hair contamination, each species hair sample was placed into a corresponding labeled glass vial or small plastic Ziploc bag.

A clear nail polish solution of Sally Hansen 'Hard as Nails' Xtreme Wear 100 Invisible was applied by brush to Fisher Scientific glass slides (see Figure 1). Animal specimen hairs were isolated by hand and carefully applied onto surface. The medial portions of the hairs were pressed into the nail polish solution at the semi-dry or 'tacky' stage of the drying process. The solution was then further air dried for 15 minutes before the embedded hairs were gently peeled away from the dried-out solution (see Figure 2).

The impressions left in the nail polish solution were kept on the glass slide. The hair slides were placed onto a Leitz HM-LUX 3 light microscope and observed at 50× and 200× magnification, respectively. Once the hair scale sample image was in range and in focus under the lens, a microscope camera – the Bodelin ProScope HR – was used to obtain the final images used.

The generalized structure of mammal hair – from the inside layer out – includes the medulla, the cortex and the cuticle (see Figures 3, 4 and 5). The cuticle is a translucent outer layer of the hair strand and consists of cuticular scales – the key identification feature for this guide.

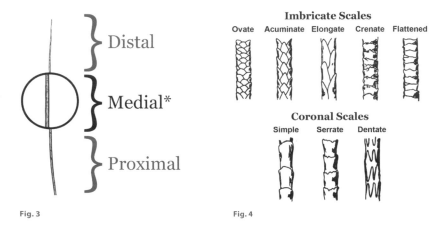

Fig. 3
Fig. 4

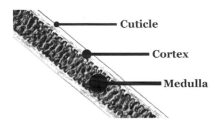

Fig. 5

Fig. 3: Anatomy of a hair, showing the three distinct areas. The medial part (or middle) of the hair was used for sample images in this guide.

Fig. 4: Key features of hair scales, adapted from Adjoran & Kolenosky (1969).

Fig. 5: Anatomy of a hair cross-section, adapted from Carrlee & Horelick (2011).

Although all species in the order Carnivora in this guide do not feature coronal scales, it is important to note that this scale feature is found in other orders of mammals. An example is the snowshoe hare (*Lepus americanus*), Figure 6, which clearly shows coronal scale traits in the guard hair. Coronal scales do not overlap, unlike imbricate scales – see moose (*Alces alces*) in Figure 7.

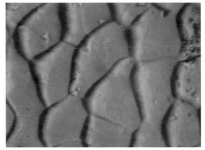

Fig. 6 (left): Close-up view (200x) of a snowshoe hare guard hair. Notice the coronal, dentate scales.

Fig. 7 (right): Close-up view (200x) of a moose guard hair. Notice the imbricate, ovate scales.

WINTER WHITE COLOUR PHASES

In some cases, certain species exhibit a complete pelage transformation. Of the 25 mammalian carnivores represented in this guide, the three species of weasel moult and change their brown summer coats for white winter coats: the short-tailed weasel (*Mustela ermina*), long-tailed weasel (*Mustela frenata*) and least weasel (*Mustela nivalis*). This adaptation camouflages these animals from predators and helps them to approach prey undetected while they hunt.

UNDERSIDES: COLOUR DIFFERENCES IN GUARD HAIRS

Certain species of mammalian carnivores exhibit a sharp contrast in guard hair colour on their underside. For these species, their hair can be consistently lighter underneath. The red fox (*Vulpes vulpes*), for instance, has a light, creamy underside in contrast to its reddish, rufous back and flanks (see page 22).

COLOUR MORPHS AND REGIONAL VARIATIONS

Although there are degrees of slight pelage variation that occur between individuals in species, certain species exhibit considerable colour morph differences or distinct pelage markings (based on geographic range).

A red fox (*Vulpes vulpes*) looks on in an open meadow for a potential meal.

CANIDAE

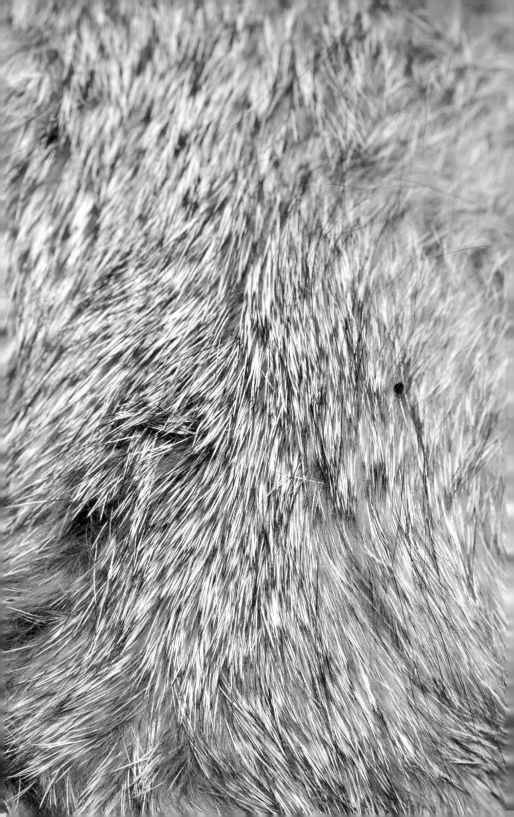

Swift Fox
Vulpes velox

Family: Canidae

Scale type: imbricate, crenate

Gross description: long, wavy, banded

Key characteristics: widely spaced scales

Similar species: coyote, grizzly bear

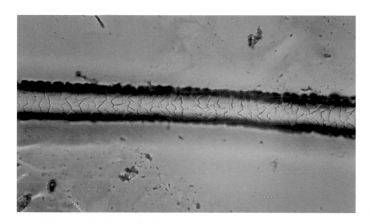

50x

200x

Arctic Fox
Alopex lagopus

Family: Canidae

Scale type: imbricate, acuminate

Gross description: long, straight, white

Key characteristics: smooth, widely spaced scales

Similar species: polar bear

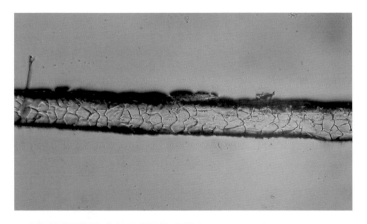

50x

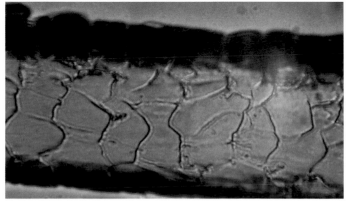

200x

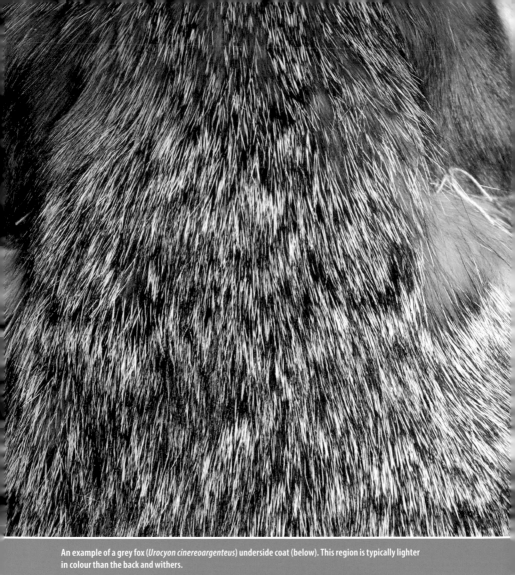

An example of a grey fox (*Urocyon cinereoargenteus*) underside coat (below). This region is typically lighter in colour than the back and withers.

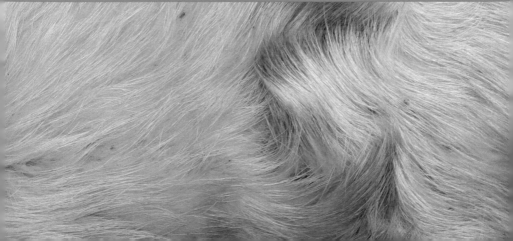

Grey Fox
Urocyon cinereoargenteus

Family: Canidae

Scale type: imbricate, crenate

Gross description: medium length, straight, banded

Key characteristics: closely overlapping, streaked scales

Similar species: American badger, raccoon, red fox

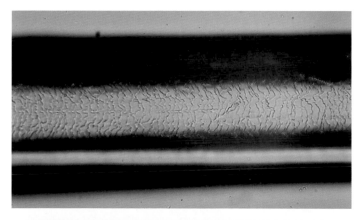

50x

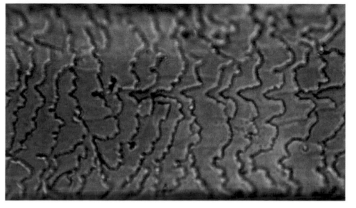

200x

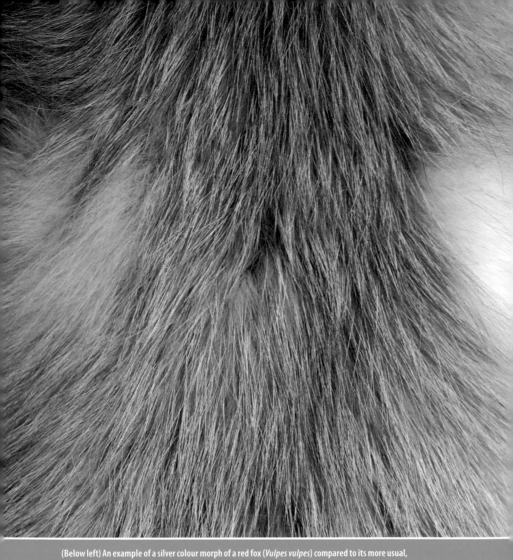

(Below left) An example of a silver colour morph of a red fox (*Vulpes vulpes*) compared to its more usual, reddish-orange, counterpart (above). (Below right) An example of its underside coat.

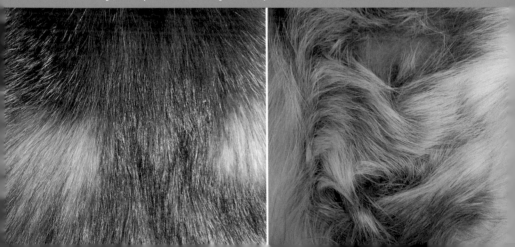

Red Fox
Vulpes vulpes

Family: Canidae

Scale type: imbricate, crenate

Gross description: long, straight, rufous/banded

Key characteristics: closely overlapping scales

Similar species: coyote, grey fox, grey wolf, raccoon

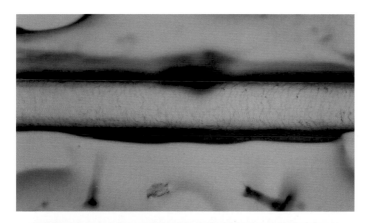

50x

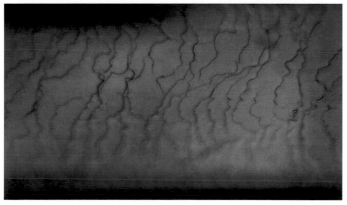

200x

Coyotes (*Canis latrans*) generally have a lighter coloured undercoat (below).

Coyote
Canis latrans

Family: Canidae

Scale type: imbricate, crenate

Gross description: long, slightly wavy, light/banded

Key characteristics: closely overlapping scales

Similar species: American badger, grey wolf, red fox

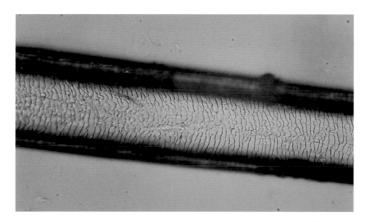

50x

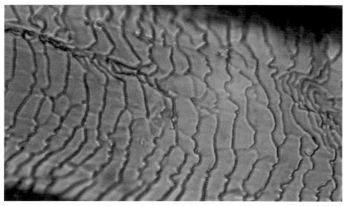

200x

Grey Wolf
Canis lupus

Family: Canidae

Scale type: imbricate, crenate

Gross description: medium/long, slightly wavy, light

Key characteristics: quite closely overlapping scales

Similar species: coyote, red fox

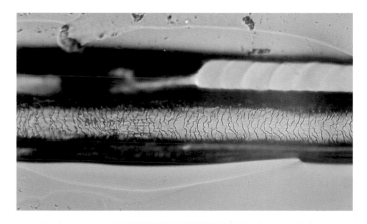

50x

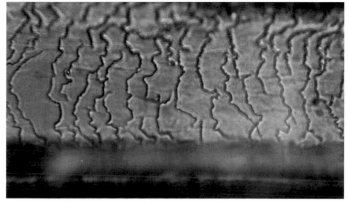

200x

A Canada lynx (*Lynx canadensis*) traverses the wintry landscape with its snowshoe-like paws.

FELIDAE

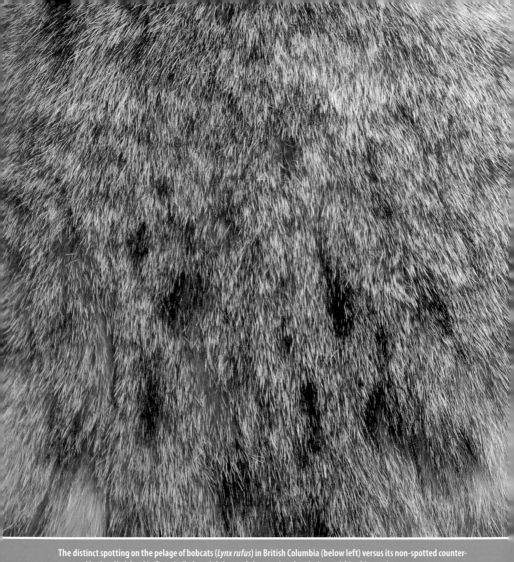

The distinct spotting on the pelage of bobcats (*Lynx rufus*) in British Columbia (below left) versus its non-spotted counterpart on Manitoulin Island in Ontario (below right) reveal clear regional variations within this species.

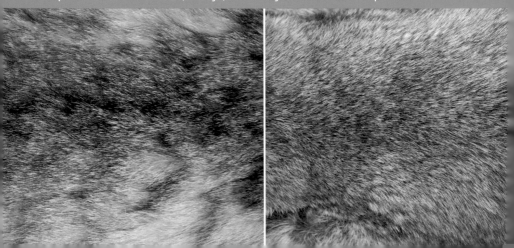

Bobcat
Lynx rufus

Family: Felidae

Scale type: imbricate, flattened

Gross description: long, straight, banded/rufous

Key characteristics: quite widely spaced scales

Similar species: Canada lynx, mountain lion

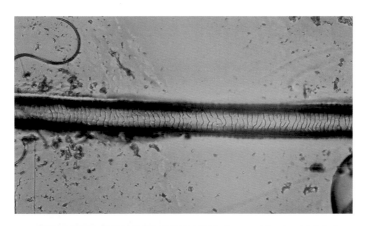

50x

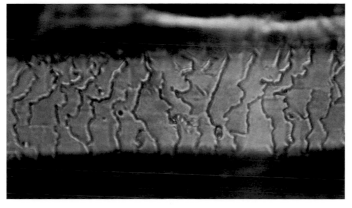

200x

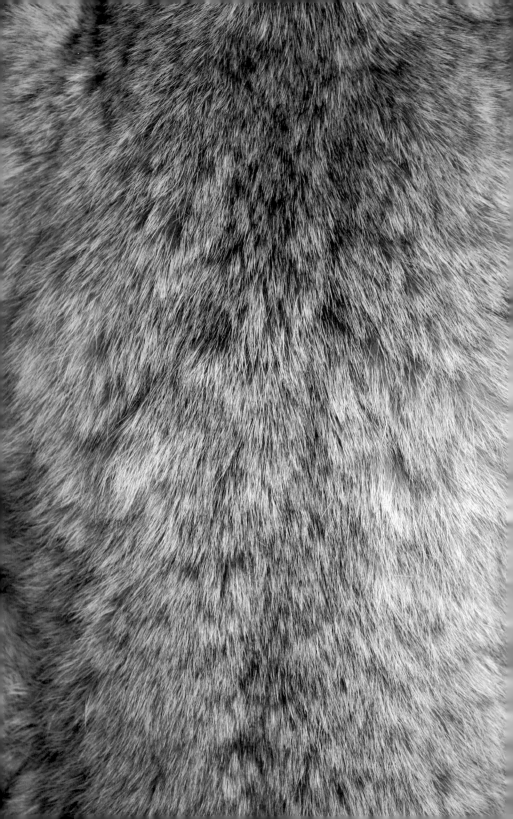

Canada Lynx
Lynx canadensis

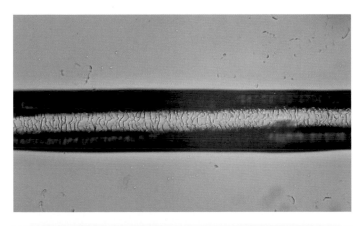

Family: Felidae

Scale type: imbricate, flattened

Gross description: long, straight, light

Key characteristics: widely spaced scales

Similar species: bobcat, mountain lion

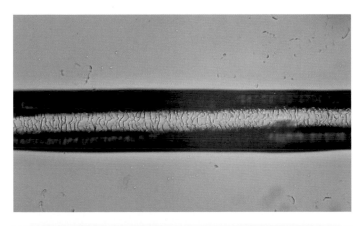

50x

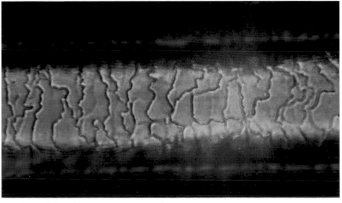

200x

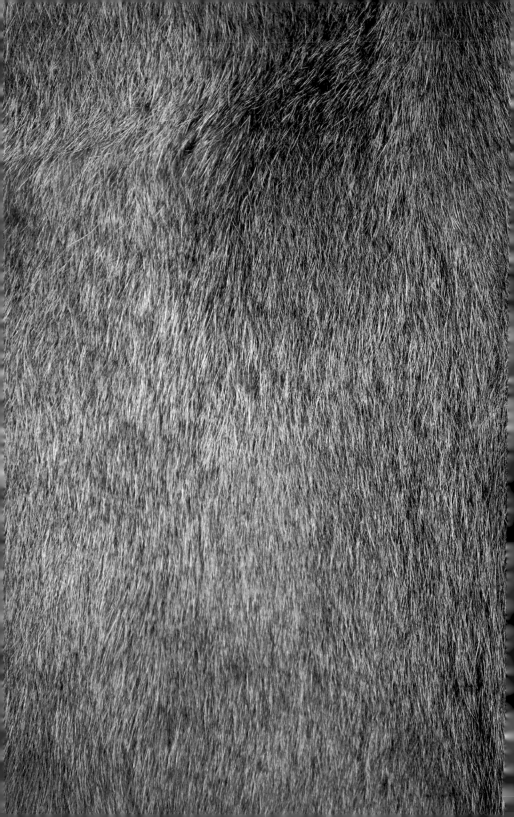

Mountain Lion
Puma concolor

Family: Felidae

Scale type: imbricate, crenate

Gross description: medium length, straight, brown/banded

Key characteristics: widely spaced, irregular scales

Similar species: bobcat, Canada lynx

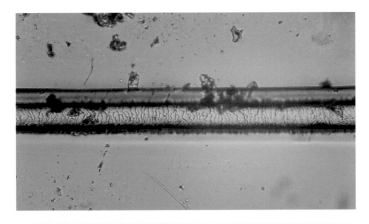

50x

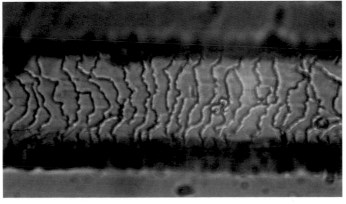

200x

An opportunistic feeder, a striped skunk (*Mephitis mephitis*) forages for insects on a log.

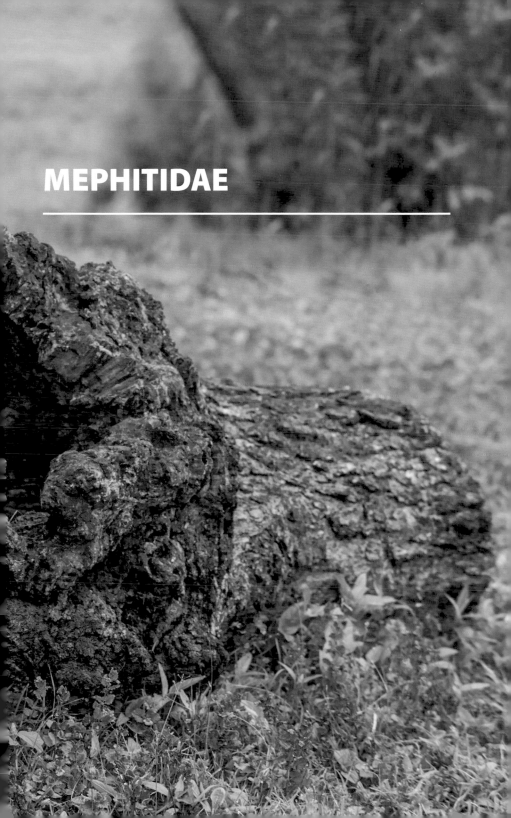

MEPHITIDAE

Western Spotted Skunk

Spilogale gracilis

Family: Mephitidae

Scale type: imbricate, crenate

Gross description: medium length, slightly wavy, black/white

Key characteristics: closely overlapping, streaked scales

Similar species: striped skunk

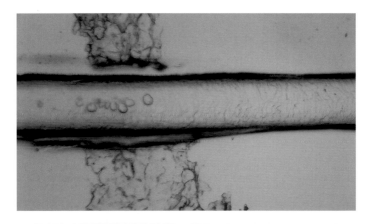

50x

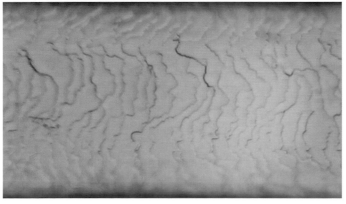

200x

Striped Skunk
Mephitis mephitis

Family: Mephitidae

Scale type: imbricate, crenate

Gross description: medium length, straight, black/white

Key characteristics: closely overlapping scales

Similar species: western spotted skunk

50x

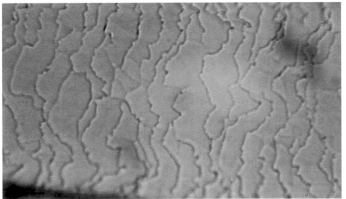

200x

An American marten (*Martes americana*) hangs about in a dense, mixed-wood forest.

MUSTELIDAE

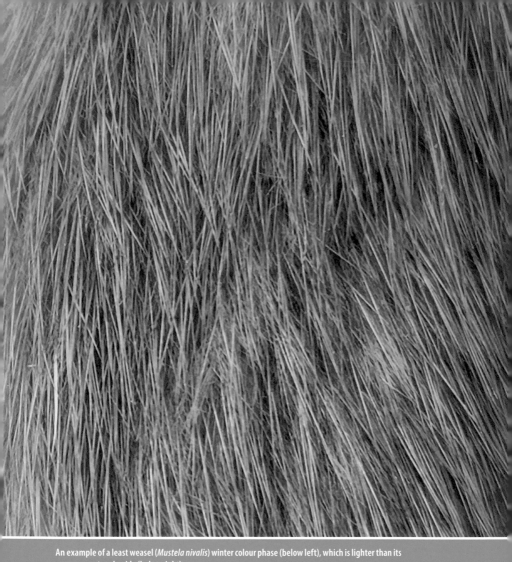

An example of a least weasel (*Mustela nivalis*) winter colour phase (below left), which is lighter than its summer coat underside (below right).

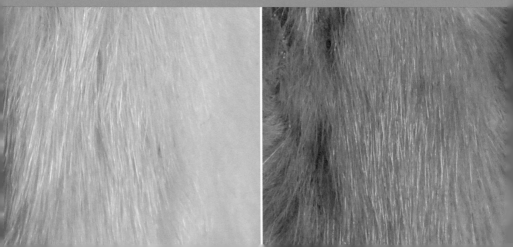

Least Weasel
Mustela nivalis

Family: Mustelidae

Scale type: imbricate, crenate

Gross description: very short, straight, white/dark

Key characteristics: closely overlapping, streaked, scales

Similar species: American mink, long-tailed weasel, short-tailed weasel

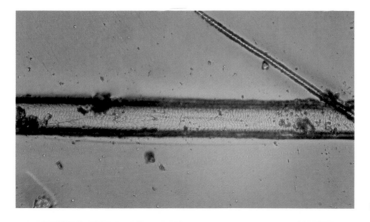

50x

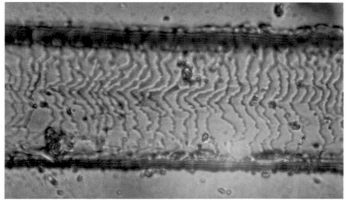

200x

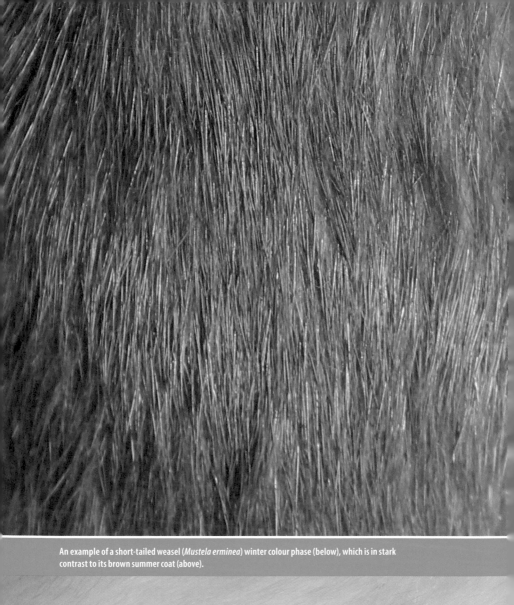

An example of a short-tailed weasel (*Mustela erminea*) winter colour phase (below), which is in stark contrast to its brown summer coat (above).

Short-Tailed Weasel
Mustela erminea

Family: Mustelidae

Scale type: imbricate, crenate

Gross description: medium length, straight, white/brown

Key characteristics: closely overlapping scales

Similar species: American mink, least weasel, long-tailed weasel

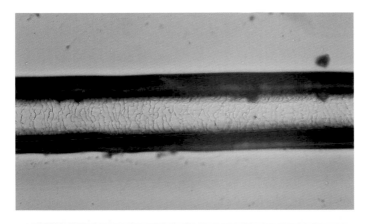

50x

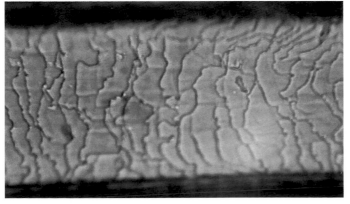

200x

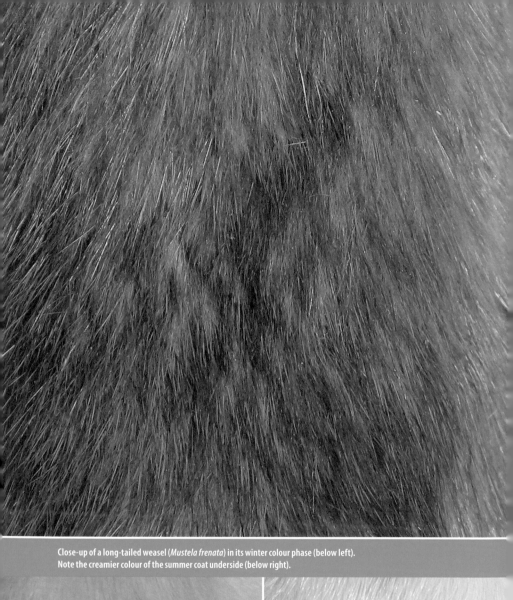

Close-up of a long-tailed weasel (*Mustela frenata*) in its winter colour phase (below left). Note the creamier colour of the summer coat underside (below right).

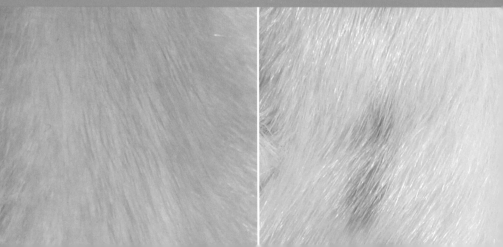

Long-Tailed Weasel
Mustela frenata

Family: Mustelidae

Scale type: imbricate, crenate

Gross description: medium length, straight, brown/white

Key characteristics: very closely overlapping, streaked scales

Similar species: American mink, least weasel, short-tailed weasel

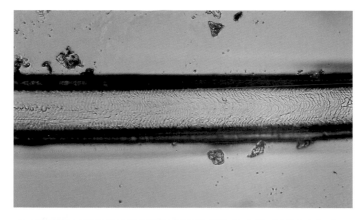

50x

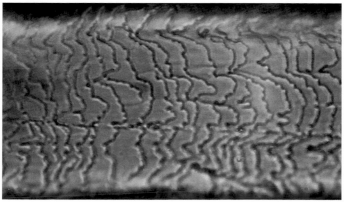

200x

American Mink

Neovison vison

Family: Mustelidae

Scale type: imbricate, crenate

Gross description: medium length, straight, brown

Key characteristics: closely overlapping, streaked scales

Similar species: American marten, fisher, least weasel, long-tailed weasel, short-tailed weasel, wolverine

50x

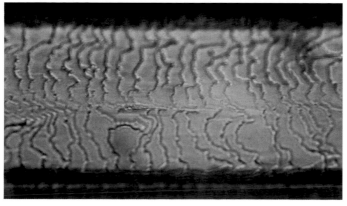

200x

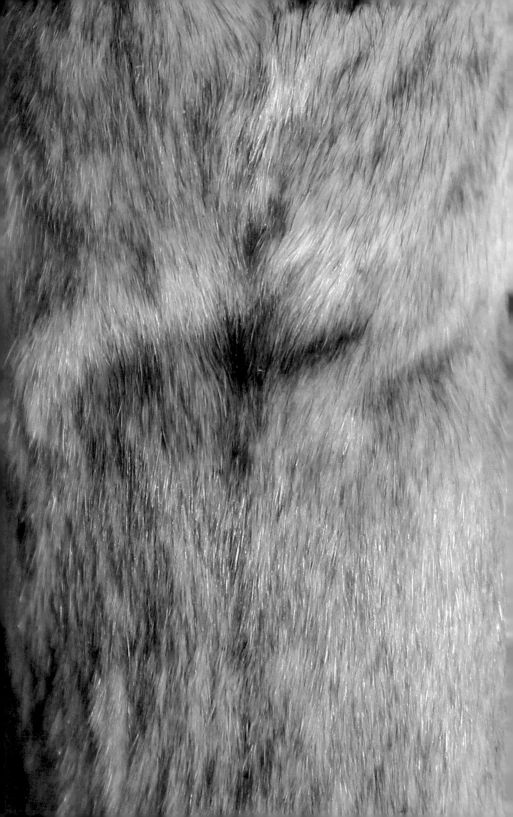

Black-Footed Ferret
Mustela nigripes

Family: Mustelidae

Scale type: imbricate, crenate

Gross description: short, straight, light

Key characteristics: closely overlapping, uneven scales

Similar species: raccoon

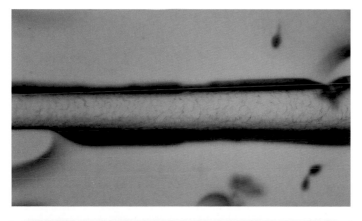

50x

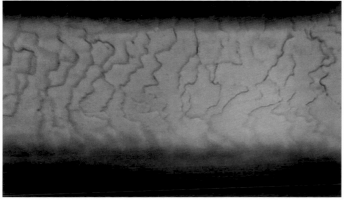

200x

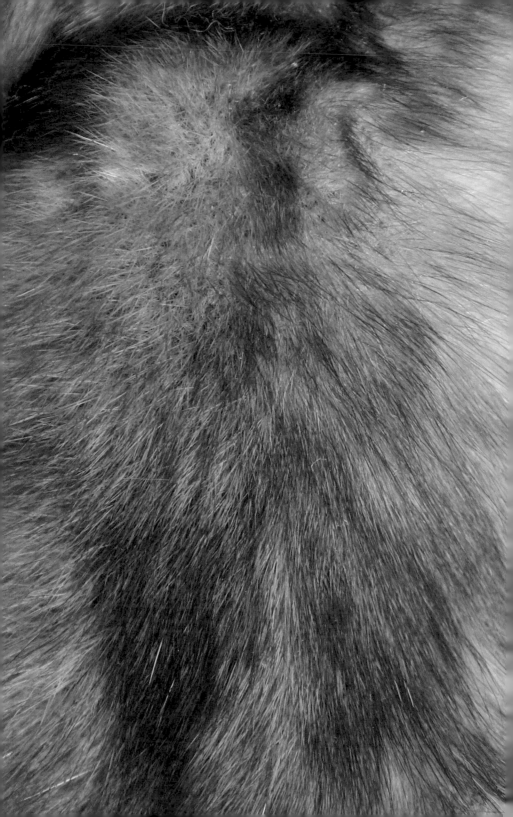

American Marten
Martes americana

Family: Mustelidae

Scale type: imbricate, crenate

Gross description: short, straight, light brown

Key characteristics: closely overlapping, streaked scales

Similar species: American mink, fisher, least weasel, long-tailed weasel, short-tailed weasel, wolverine

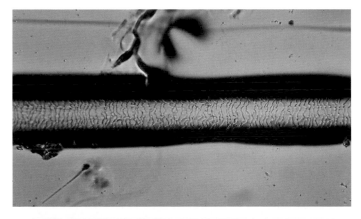

50x

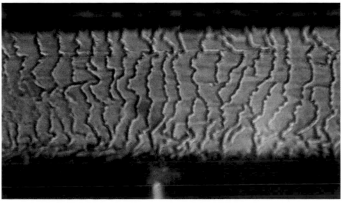

200x

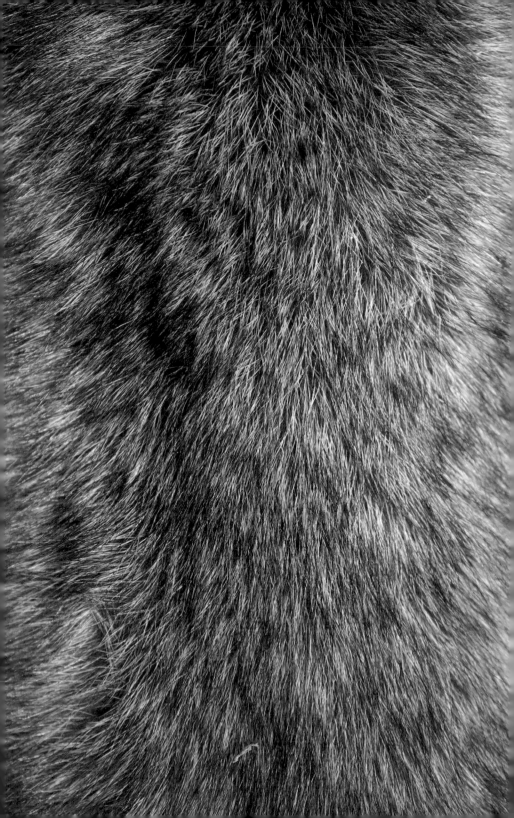

Fisher

Pekania pennanti

Family: Mustelidae

Scale type: imbricate, crenate

Gross description: long, slightly wavy, dark

Key characteristics: closely overlapping scales

Similar species: American mink, northern river otter, short-tailed weasel, wolverine

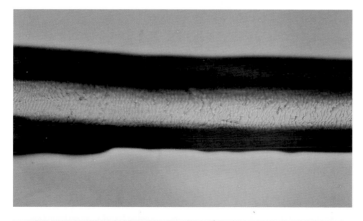

50x

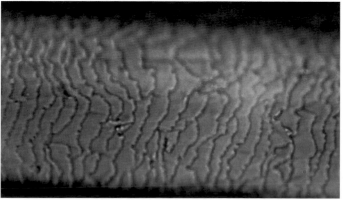

200x

American Badger
Taxidea taxus

Family: Mustelidae

Scale type: imbricate, crenate

Gross description: medium length, slightly wavy, banded

Key characteristics: closely overlapping, streaked scales

Similar species: coyote, grey fox, raccoon

50x

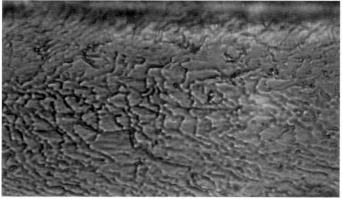

200x

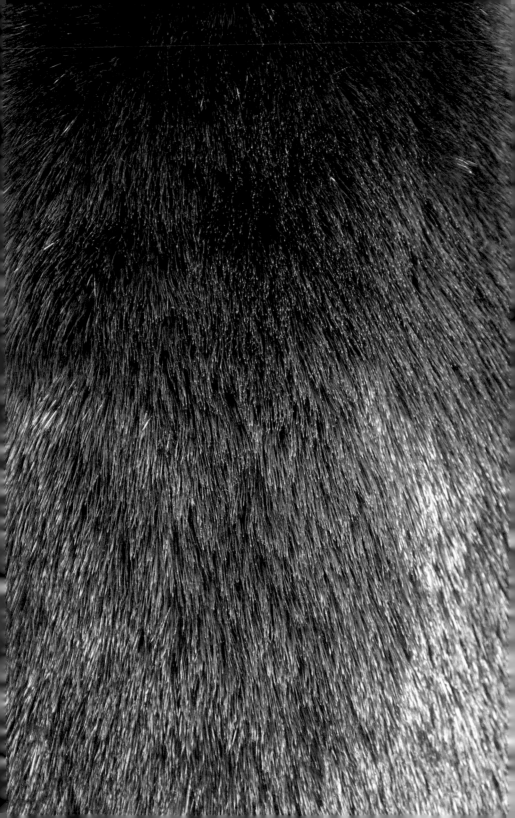

Northern River Otter
Lontra canadensis

Family: Mustelidae

Scale type: imbricate, crenate

Gross description: medium length, straight, dark

Key characteristics: closely overlapping, irregular scales

Similar species: American mink, fisher, long-tailed weasel, short-tailed weasel

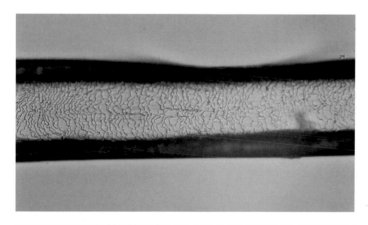

50x

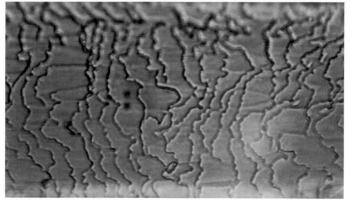

200x

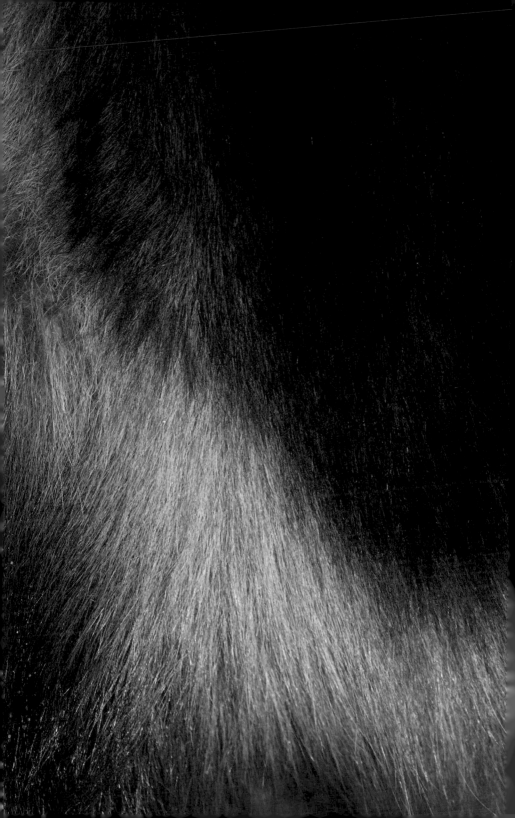

Wolverine
Gulo gulo

Family: Mustelidae

Scale type: imbricate, crenate

Gross description: long, straight, brown/dark

Key characteristics: very closely overlapping, thick, streaked scales

Similar species: American mink, fisher, northern river otter

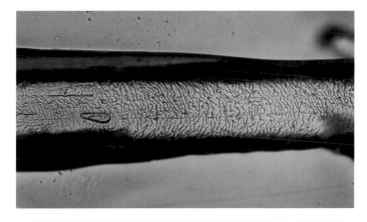

50x

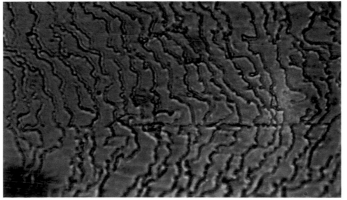

200x

A raccoon (*Procyon lotor*) cautiously observes its surroundings from the safety of a cedar tree.

PROCYONIDAE

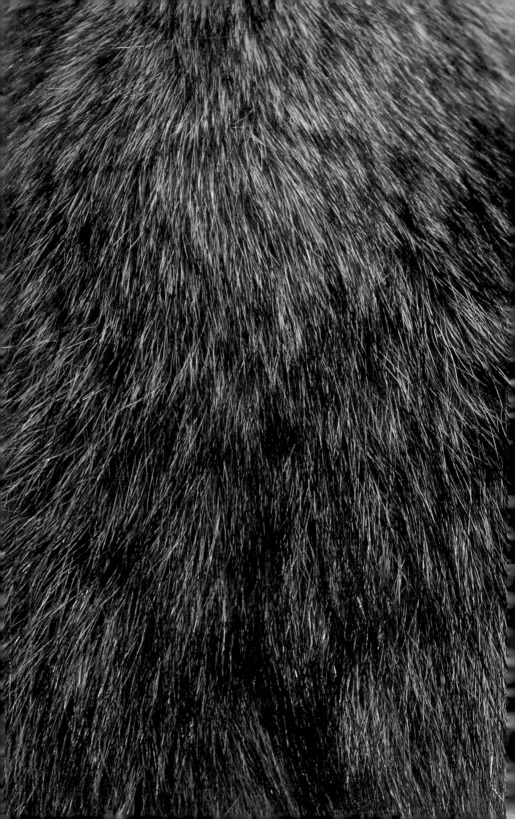

Raccoon
Procyon lotor

Family: Procyonidae

Scale type: imbricate, crenate

Gross description: medium/long, straight, banded

Key characteristics: closely overlapping scales

Similar species: coyote, grey fox, grey wolf, red fox

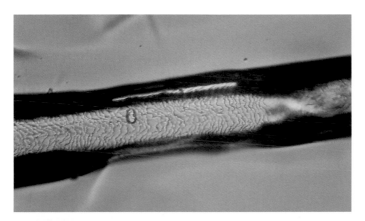

50x

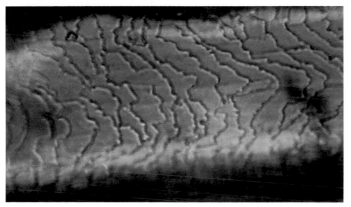

200x

The king of the Arctic, and largest land carnivore on Earth, a polar bear (*Ursus maritimus*) scans the horizon for prey.

URSIDAE

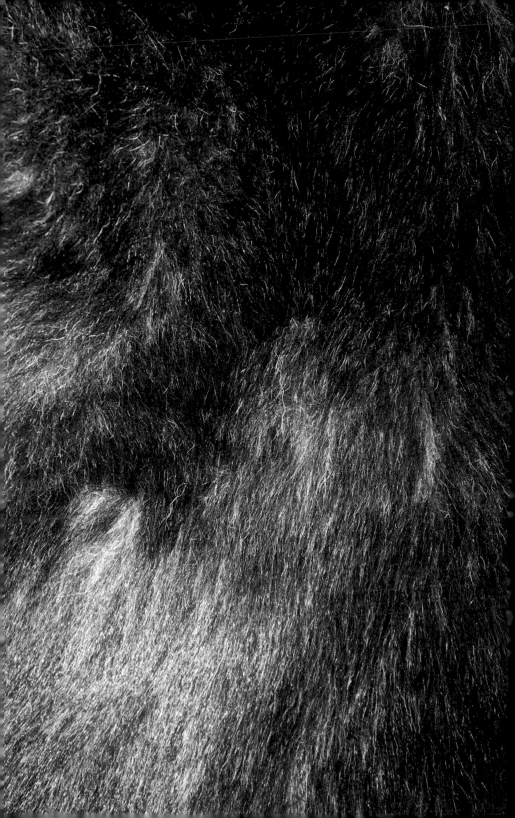

Black Bear
Ursus americanus

Family: Ursidae

Scale type: imbricate, crenate

Gross description: long, wavy, dark

Key characteristics: quite widely spaced scales

Similar species: grizzly bear

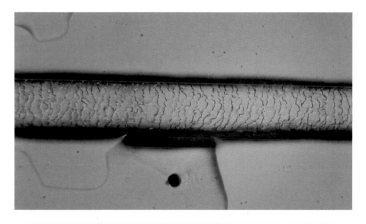

50x

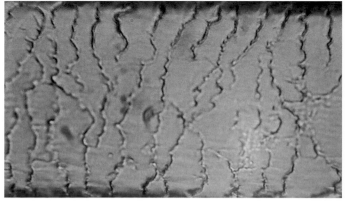

200x

Grizzly Bear
Ursus arctos

Family: Ursidae

Scale type: imbricate, crenate

Gross description: medium length, wavy, brown

Key characteristics: very widely spaced scales

Similar species: black bear, swift fox

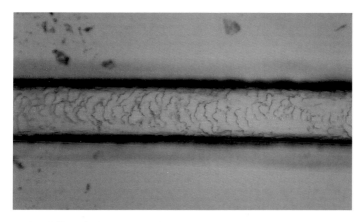

50x

200x

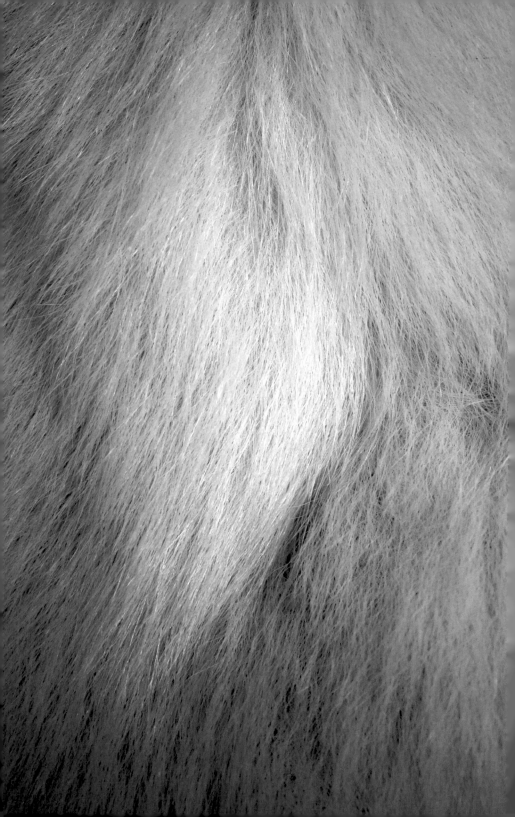

Polar Bear
Ursus maritimus

Family: Ursidae

Scale type: imbricate, crenate

Gross description: medium length, slightly wavy, white

Key characteristics: irregular/widely spaced scales

Similar species: Arctic fox

50x

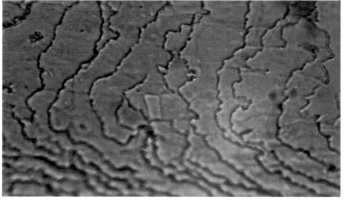

200x

References

Adorjan, A. and G. Kolenosky (1969). *A Manual for the Identification of Hairs of Selected Ontario Mammals*. Research Branch – Ontario Department of Lands and Forests. Monograph. BCIN No. 136294.

Carrlee, E. and L. Horelick (2011). The Alaska fur ID project: a virtual resource for material identification. AIC Objects Specialty Group Postprints 18: 149–171. https://alaskafurid.wordpress.com/

Eder, T. and G. Kennedy (2011). *A Manual for the Identification of Hairs of Selected Ontario Mammals* (UK edition). Lone Pine Publishing.

Hausman, L. A. (1924). Further studies of the relationships of the structural characters of mammalian hair. *The American Naturalist* 58 (659): 544–557. https://doi.org/10.1086/280006

Naughton, D. and C. M. of Nature (2012). *The Natural History of Canadian Mammals*. University of Toronto Press, Scholarly Publishing Division. https://doi.org/10.3138/9781442669574

Teerink, B. J. (1991). *Hair of West European Mammals: Atlas and Identification Key*. Cambridge University Press.

Appendix 1: A Summary of Scale Types

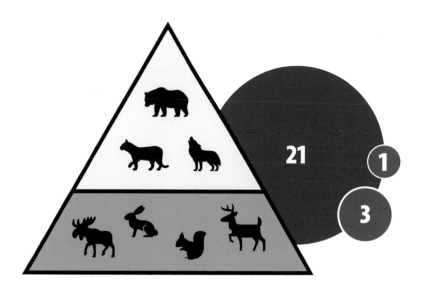

All 25 mammalian carnivore species of Canada have *imbricate* scales.

Of those, 21 have *crenate* edges; three have *flattened* edges; and one species (Arctic fox) has *smooth/acuminate* edges (see Figure 8).

All three felidae species feature *broken*, *flattened*-edge type scales.

Some carnivore families exhibit certain hair scale characteristics, such as deeply serrated and tight scales (Mustelidae).

Species in the order herbivora (green above) display various scale and edge types but *do not appear to ever have crenate hair scales.*

(Icons taken from https://www.iconshock.com/iphone-icons/animals-icons/)

Appendix 2: Patterns and Comparisons

....

Patterns observed

MUSTELIDAE

Key characteristics: deeply serrated, tight scales; wavy and more compact than other families in order Carnivora.

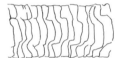 

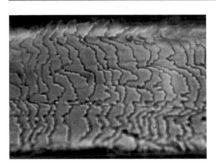

Long-Tailed Weasel
Mustela frenata

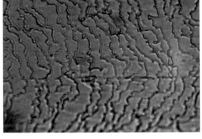

Wolverine
Gulo gulo

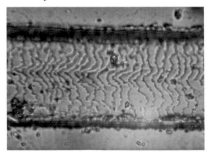

Least Weasel
Mustela nivalis

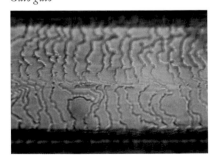

American Mink
Neovison vison

FELIDAE

Key characteristics: scales are typically more broken and flattened; width of a scale is approximately one tenth the diameter of the guard hair.

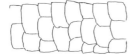 

Bobcat
Lynx rufus

Canada Lynx
Lynx canadensis

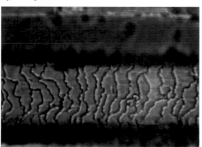

Mountain Lion
Puma concolor

Appendix 2: Patterns and Comparisons cont.

Comparisons

FELIDAE    **MUSTELIDAE**

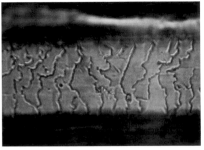

Bobcat
Lynx rufus

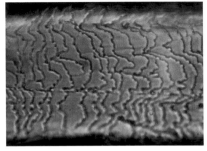

Long-Tailed Weasel
Mustela frenata

Canada Lynx
Lynx canadensis

American Mink
Neovison vison